SHOCK WAVES

THROUGH LOS ANGELES

THE NORTHRIDGE EARTHQUAKE

Carole Garbuny Vogel

LITTLE, BROWN AND COMPANY

Boston New York Toronto London

To my uncle, Professor Siegfried Garbuny, for always being there

Acknowledgments

I would like to thank the following people for their technical assistance: Michael J. Rymer, geologist, U.S. Geological Survey, Menlo Park, for his insight and invaluable manuscript critique; Nick Delli Quadri, senior structural engineer, City of Los Angeles Department of Building and Safety, for his ability to communicate complex engineering principles in laymen's terms; Mary Edwards, information and planning chief, Northridge Earthquake Long-Term Recovery Area Office, FEMA, for supplying information and critiquing the manuscript; and Judy Steele, legislative analyst, City of Los Angeles, for reviewing the manuscript.

I am grateful to the following individuals who took the time to answer my questions in depth and provide me with data: Captain Louis Casas, Community Response Team North, Disaster Preparedness Section, Los Angeles City Fire Department; Jim Mori, scientist-in-charge, Pasadena Office, U.S. Geological Survey; Pat Bonino, manager of the Disaster Recovery Team, City of Los Angeles; Valerie Shane of FEMA; and Sue Pitts of the Office of Media Relations at the California Institute of Technology, Pasadena.

Special thanks to Sharon Begley of *Newsweek* magazine for permission to use her plate tectonic analogy that compares the earth's structure to a chocolate-covered cherry, as well as to Francine Della Catena of the *Los Angeles Times* and Gary V. Phelps of the *Ventura County Star* for their help in procuring photographs. I would also like to acknowledge the help of the dozens of other people who took time from their busy schedules to answer my questions. Where the opinions of experts conflicted, I used my own judgment in presenting the information.

I am especially indebted to fellow writer Florence Harris and my husband, Mark A. Vogel, for their excellent advice and support, to my nephew, Daniel Butterworth, for providing the student perspective, and to Coralee Paull for her capable assistance with research. As always, my agent, Renée Cho of McIntosh & Otis, and my editor, Hilary Breed, have managed to make the work of writing fun.

First Edition

Photography and illustration credits: © Sean Donovan: diagrams, p. 31. The Earthquake Engineering Research Center, University of California at Berkeley: pp. 4, 15, 25 left, 26, 27 top and bottom. © ***Los Angeles Times:*** by Lacy Atkins, p. 13; by Rod Boren, p. 6 top; by Gerard Burkhart, p. 11; by Brian Vander Brug, p. 21 top; by Jill Connelly, p. 12 top; by Michael Edwards, pp. 5 top and 10; by Gary Friedman, p. 22 bottom; by Ken Lubas, pp. 8 top and 14; by Joel P. Lugavere, title page; by Julie Markes, pp. 3, 6 bottom, and 23 right; by Anacleto Rapping, p. 21 bottom; by Perry C. Riddle, pp. 5 bottom and 25 right; by George Wilhelm, p. 22 top. U.S. Geological Survey: p. 8 bottom; by Michael J. Rymer, p. 29; by Robert E. Wallace, p. 28. © ***Ventura County Star:*** by K. C. Alfred, p. 19 bottom; by Gary V. Phelps, p. 32; by Mark Pickering, pp. 12 bottom, 18, and 24 top; by Kevin Rice, p. 24 bottom; by Walter Thompson, p. 23 left; by Scott Weersing, pp. 7, 19 top, and 20. © Mark A. Vogel: maps, pp. 9, 16–17, and 130, based on maps published by the State of California Division of Mines and Geology.

Library of Congress Cataloging-in-Publication Data

Vogel, Carole Garbuny.
Shock waves through Los Angeles : the Northridge earthquake / by Carole Garbuny Vogel. — 1st ed.
p. cm.
Summary: Examines the causes of the earthquake that hit the Northridge area of southern California in 1994 and describes the devastating effects of the quake.
ISBN 0-316-90240-3
1. Earthquakes — California — Northridge (Los Angeles) — Juvenile literature. 2. Buildings — Earthquake effects — California — Northridge (Los Angeles) — Juvenile literature. [1. Earthquakes — California — Northridge (Los Angeles)] I. Title.
QE535.2.U6V64 1996
363.3'495 — dc20 95-52104

10 9 8 7 6 5 4 3 2 1

Published simultaneously in Canada
by Little, Brown & Company (Canada) Limited

Printed in China

This home in Sherman Oaks, California, plunged down a hill.

AT 4:31 A.M. ON JANUARY 17, 1994, a powerful earthquake jolted millions of people awake in southern California. The quake damaged more than 112,000 buildings, buckled three major freeways, and sparked scores of fires. Fifty-eight people were killed, hundreds were critically injured, and thousands were left homeless. Centered in the San Fernando Valley near Northridge, a community only 20 miles northwest of downtown Los Angeles, this event became known as the 1994 Northridge earthquake.

The earthquake — also known as a temblor — began 12 miles underground, beneath the San Fernando Valley. An immense slab of rock suddenly shifted, heaving the mountains north of Northridge more than a foot upward. In an instant, shock waves radiated out in all directions. They struck first at the epicenter, the area directly above the origin of the quake. Traveling through the earth at a rate of two to three miles per second, the shock waves were felt as far away as Las Vegas, 275 miles to the northeast, San Diego, 125 miles to the southeast, and on into Mexico. The actual lurching of the earth took about eight seconds. But farther away from the quake's origin, the shaking lasted up to 45 terrifying seconds. Because of its proximity to the epicenter, the San Fernando Valley, home to three million people, suffered the brunt of the quake.

The temblor collapsed seven freeway bridges, affecting four major highways. Reporting to duty after the quake struck, police officer Clarence W. Dean unknowingly sped up a severed freeway ramp on his motorcycle. He flew over the edge to his death 25 feet below and became one of three fatalities in the dozens of quake-related traffic accidents.

The overall casualty count from the temblor could have been much higher than the 58 reported deaths. Had the quake occurred when the freeways were jammed and the shopping areas and parking structures were crowded, thousands more people might have perished. But the earthquake struck before dawn, when the highways and public buildings lay nearly deserted.

The failure of this overpass cost police officer Clarence W. Dean his life.

The quake stranded these vehicles on the Golden State Freeway (*above*).
The driver of this car survived the collapse of an overpass on the Simi Valley Freeway (*below*).

Luckily no sports fans were in Anaheim Stadium when the giant TV screen and billboards toppled onto the upper deck.

Some phones still worked the morning of the quake, enabling people to contact friends and relatives outside the quake zone.

Most southern Californians were sleeping when the earthquake began. They awoke to a loud rumbling and to houses that shook like bucking broncos. Windows and mirrors shattered, books flew from shelves, and furniture careened across rooms. Many residents experienced two savage jolts separated by a roller-coaster motion. In the area of the epicenter, the vibration was so powerful, it was nearly impossible to stand or walk. Couples clutched each other in terror. Frightened children screamed for their parents.

After what seemed like eternity, the fierce rocking finally stopped. Petrified residents rushed to check that others in their homes were safe. With no light and often no shoes, many cut their feet on shards of glass. Worried that the ceilings might collapse, people stumbled outdoors in the dark, clad in pajamas or wrapped in sheets. Every few minutes, aftershocks rippled through the ground, compounding the fear.

In the glow of flashlights and headlight beams, emergency crews quickly began to assess the losses. Most modern steel-frame high-rise buildings appeared unscathed on the outside. However, older brick buildings and some newer wood-frame apartment houses with ground-floor garages sustained tremendous damage. The worst destruction occurred in structures built prior to the enactment of tough building codes in the 1970s. These codes were adopted after a temblor in 1971 damaged more than 30,000 buildings in the San Fernando Valley. If these strict building codes had not been in place during the two decades preceding the Northridge earthquake, the destruction wrought by this temblor would have been catastrophic.

In Fillmore, the exterior walls of the Fillmore Hotel peeled away, crushing several cars parked below. Farmworkers and their families were occupying these rooms on the hotel's second floor when the quake struck.

Los Angeles (*top*) experienced more than 9,000 aftershocks in the six months following the Northridge quake. Eight of these tremors were strong enough to sway buildings.

Seismologist Jim Mori monitors earthquake activity from his office at the U.S. Geological Survey in Pasadena, California (*bottom*).

Earthquakes are nothing new to southern California. Thousands of temblors occur there each year. Most are so minor they cannot even be felt. However, every 40 years or so, a severe quake rocks the region.

The area is riddled with a vast series of faults — gigantic fractures in the earth's crust where temblors form. Until 1987, when a flurry of earthquakes occurred in the Los Angeles vicinity, scientists were not aware of how large a number of faults lay hidden beneath the populated region. They thought the main danger to the city was the threat of a major temblor along the San Andreas, an exposed fault northeast of Los Angeles.

Since then, nearly 100 faults in six major fault zones have been identified as being capable of producing a significant quake in the Los Angeles area. Despite their heightened awareness, the researchers were caught off guard by the Northridge quake. Until it struck, most scientists did not even know of the existence of the fault that caused it. As a result, they have since been scrambling to reform their understanding of the fault zones underlying the region.

The Northridge quake released nearly as much pent-up energy as the 1980 eruption of the Mount St. Helens volcano. The earthquake, like all temblors, generated shock waves that spread out at different rates. As the shock waves traveled through the earth, sensitive instruments called *seismographs* recorded the vibrations. By comparing the arrival times of different shock wave types and their intensity at various seismograph locations, scientists determined the origin of the temblor and its size.

The Northridge earthquake registered 6.7 on the *moment-magnitude scale,* a measure of earthquake energy that replaced the outdated Richter magnitude scale. This new measure takes into account the size of the fault where an earthquake occurs and the distance the earth shifts. The larger the fault and the greater the shift, the higher its moment magnitude. Earthquakes with magnitudes above five can be quite damaging, especially when they strike near or below large cities.

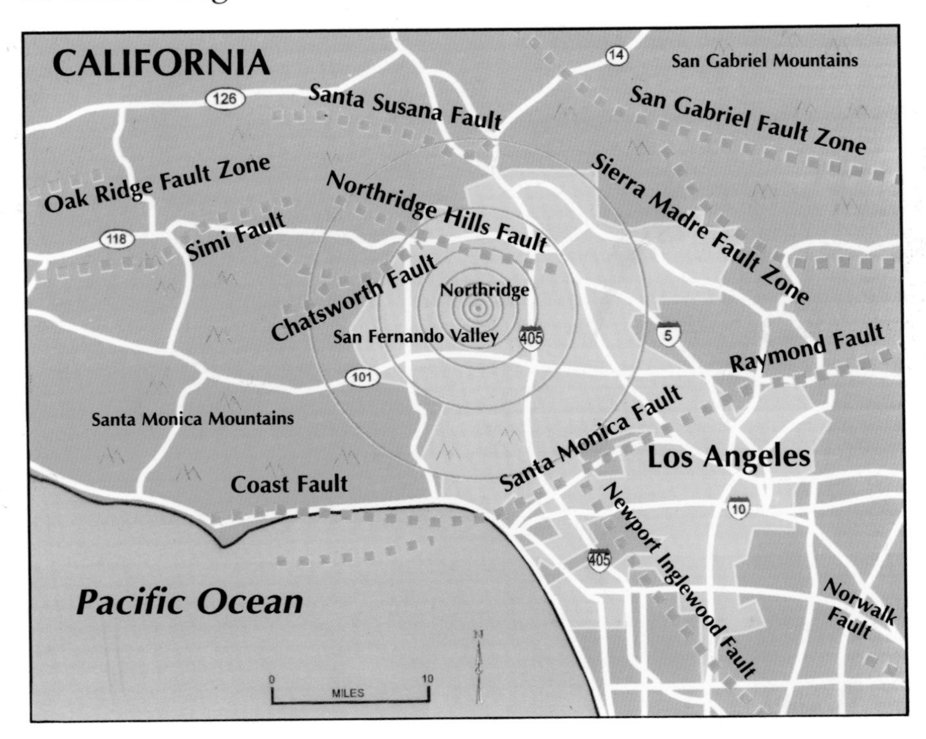

Six major fault zones underlie the Los Angeles area. The fault responsible for the Northridge earthquake is probably part of the Oak Ridge fault zone.

Noplace was harder hit during the Northridge earthquake than Northridge Meadows, a 164-unit apartment complex. It lurched eight feet, causing the top two floors to collapse onto the bottom level. Sixteen people perished in the crush, and dozens of others were trapped in the debris.

In the predawn darkness, some survivors managed to slither to safety on their own. Others were pulled out or coaxed out by neighbors. Those pinned deep in the rubble could not be reached. With each aftershock, the wreckage dropped lower and lower. Terrified, the trapped residents wondered whether they would ever get out alive.

The collapsed Northridge Meadows apartments in Northridge were the site of nearly one third of the fatalities and many heroic rescues.

Two survivors of the Northridge Meadows complex disaster, Jason Lee, age 12, and his mother, Hyun Sook Lee, attend the funeral of Jason's older brother, Howard, and father, Pil Soon Lee.

Specially trained fire department search-and-rescue crews began to arrive within an hour of the main shock. Using chain saws and other equipment to cut through the ruins, they risked their own lives to extricate fourteen trapped apartment dwellers and recover the bodies of the dead. Aftershocks periodically loosened debris and temporarily trapped some of the rescuers. Nearly 30 hours after they started, crew members pulled the last victim from the rubble.

Nearby, at the Northridge Fashion Center Mall, another team of highly skilled firefighters was struggling to save Salvador Pena. The 34-year-old maintenance worker had been sweeping the mall's parking garage when the temblor flattened the structure. Tons of fallen concrete trapped Pena inside his small sweeper truck, crushing his legs and right hand. The situation initially looked hopeless.

Laboring all day, firefighters burrowed through the rubble with jackhammers, then shored up their tunnel with wood. Piece by piece they removed chunks of debris. Pena was wedged in so tightly that the rescue crew even considered amputating his legs to release him. Fortunately air bags used to lift the concrete were sufficient to allow firefighters to pull Pena free of the wreckage.

Members of the Los Angeles County Fire Department at the site where Salvador Pena was trapped (*left*).

After working for seven hours to free Salvador Pena, the successful rescue crew whisks him away on a stretcher.

More than 600 natural gas pipelines ruptured beneath the streets. All leaked gas, and a few exploded, sending plumes of flame skyrocketing into the air. Broken gas lines within or connected to buildings also triggered fires.

Flames from an exploding pipeline torched homes along Balboa Boulevard, while water from a burst water main flowed away, useless.

Using water from a swimming pool, frantic Los Angelenos try to contain one of the 100 or so fires that broke out immediately after the first shock.

Shattered water mains erupted like geysers. They transformed some streets into rushing torrents, flooding homes and businesses nearby. With so much water draining away, faucets ran dry in two thirds of the homes in the San Fernando Valley. And water was in short supply for fire fighting. At least 600 buildings burned, including 172 mobile homes. Nevertheless, firefighters managed to extinguish all the fires within six hours using water from tankers and helicopter drops.

The temblor snapped power lines and severely damaged many electrical distribution stations. All of Los Angeles and much of the surrounding area lost electricity, depriving nearly four and a half million residents of power. The blackout forced the automatic shutdown of a power plant in Utah and triggered temporary outages in six other western states and part of Canada. Los Angelenos marveled at one outcome of the unaccustomed blackness: Without light pollution from the city, they could see a sky bursting with stars.

In some Los Angeles neighborhoods, the power failure knocked out local telephone service for about half a day. When service was reinstated, the long distance carriers blocked calls from out of state in order to keep phone lines free for emergency use. Hundreds of computer users with working phones and restored power relied on computer networks to relay messages in and out of the ravaged area.

The loss of electricity disabled the main water treatment plant responsible for purifying Los Angeles's water supply. And there was the possibility that dirt and other contaminants could enter the water through cracked pipes. So concerned health officials warned residents to boil water before drinking it.

Water-main breaks like this one shunted water away from fire hydrants.

CALIFORNIA
MAP LEGEND
HIGHWAY COLLAPSE
TRAIN DERAILMENT
FIRE
BUILDING COLLAPSE
EPICENTER
126
Santa Clara River
VENTURA COUNTY
Moorpark
118
Ventura Freeway
Santa Monica Mountains
N
0
10
MILES
Pacific Ocean

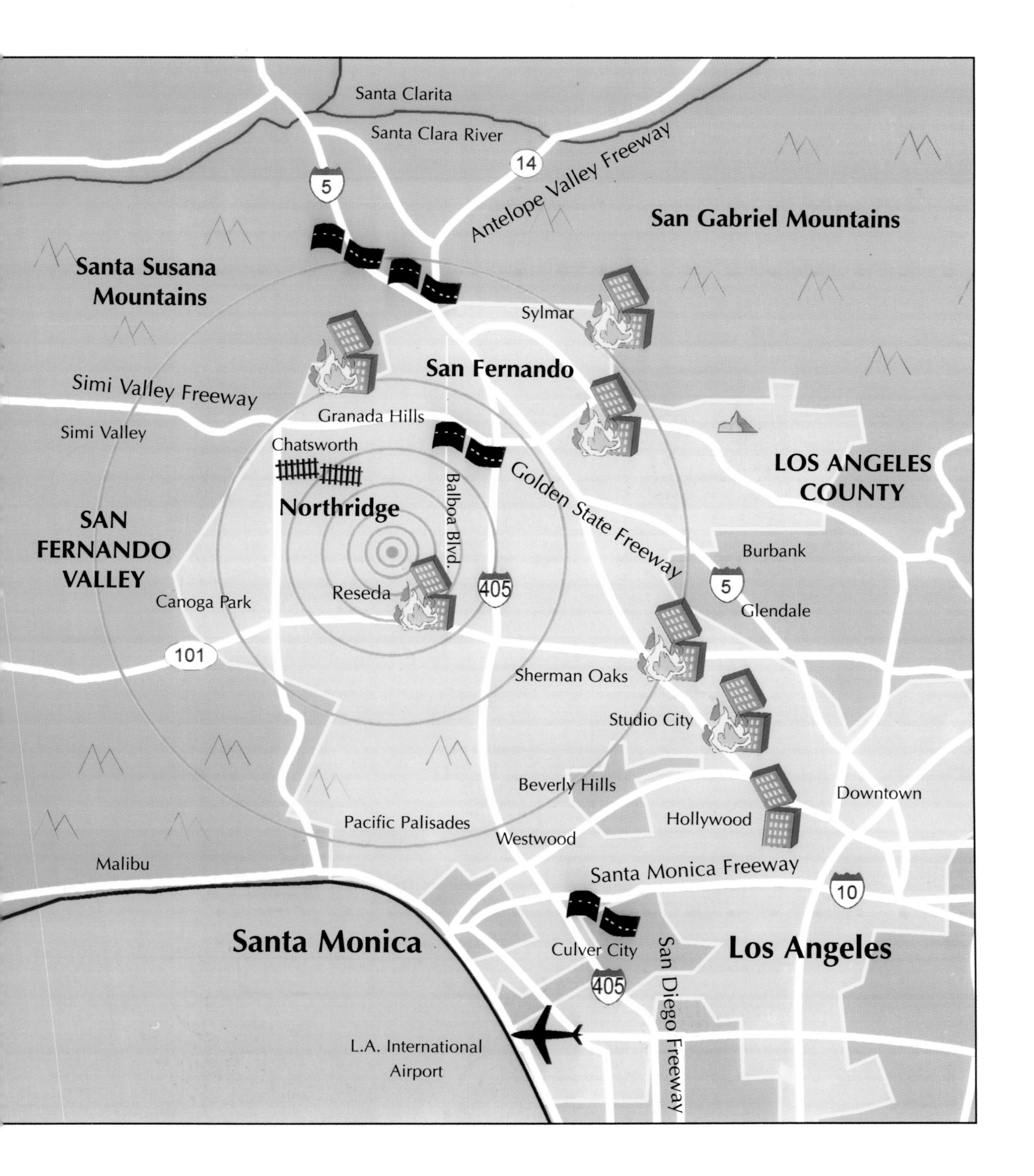

Santa Clarita
Santa Clara River
5
14
Antelope Valley Freeway
San Gabriel Mountains
Santa Susana Mountains
Sylmar
San Fernando
Simi Valley Freeway
Simi Valley
Granada Hills
Chatsworth
LOS ANGELES COUNTY
Golden State Freeway
Balboa Blvd.
Northridge
SAN FERNANDO VALLEY
Burbank
5
405
Reseda
Canoga Park
Glendale
101
Sherman Oaks
Studio City
Beverly Hills
Downtown
Hollywood
Pacific Palisades
Westwood
Malibu
Santa Monica Freeway
10
Santa Monica
Culver City
San Diego Freeway
Los Angeles
405
L.A. International Airport

Many hospitals and other medical facilities experienced heavy damage, especially to their interiors. All lost power and had to rely on backup generators for electricity. At the Granada Hills Community Hospital, a rooftop water tank flew off its hinges and spilled 2,500 gallons of water. The resulting flood necessitated the transfer of patients from the hospital's second and third floors to other parts of the building. Still, the hospital fared better than the neighboring Kaiser Permanente clinic, which was nearly demolished by the quake.

This Kaiser Permanente clinic in Granada Hills was one of five medical facilities that suffered catastrophic damage.

Rescue workers in Fillmore set up a triage center at San Cayento School to help injured quake victims.

Almost 9,000 people required medical attention in the quake zone. Swamped by the tremendous influx of patients, hospitals converted their parking lots into makeshift emergency rooms. Medical personnel treated countless broken bones and lacerations from flying debris and broken glass. They also dealt with quake-related heart attacks, as well as the usual illnesses and baby deliveries. Medical teams from distant areas rushed in to help.

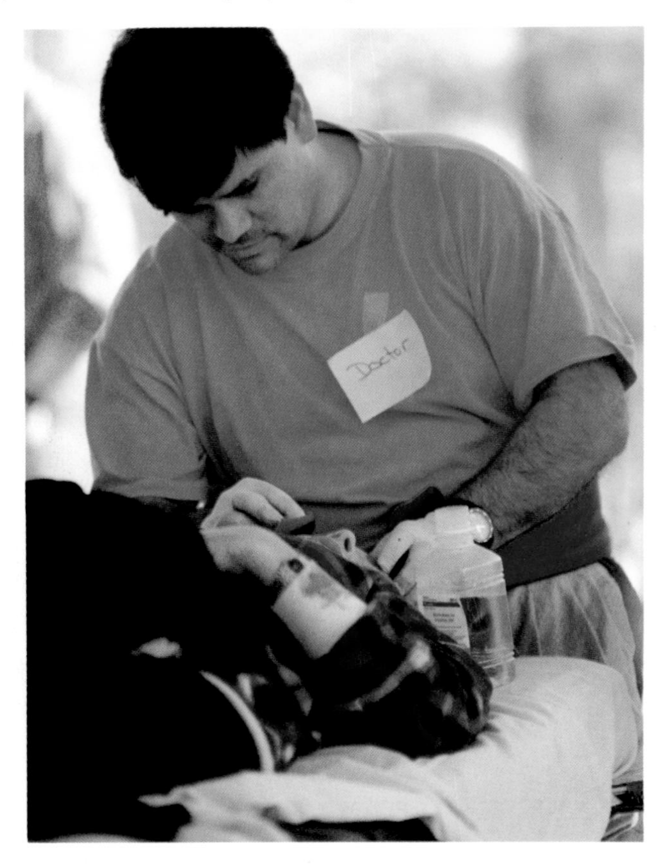

Dr. Richard Wagner treats a patient in the parking lot of Simi Valley Hospital.

Fear that aftershocks might bring down already weakened buildings kept families in Fillmore out of their homes.

In the aftermath of the quake, the stunned population of southern California tried to come to grips with their misfortune. Fearful children clung to their parents, and anxious parents kept a close watch on their offspring. Many people surveyed the chaos around them and wondered what to do next. Others lined up at stores in search of food, batteries, cleaning supplies, bottled water, and diapers.

Simply getting around was a nightmare. A glistening sea of shattered glass covered the sidewalks. Traffic lights blacked out by the power failure added to the pandemonium created by ravaged freeways and other mangled roads.

Structural engineers began to inspect damaged homes and commercial buildings. Nearly 17,000 structures were eventually deemed unsafe. Depending on the extent of the destruction, occupants of these buildings were either barred from entering or limited in the time they could spend inside. Sometimes displaced residents broke into their own homes and risked serious injury to collect clothes and other essentials. Seventeen neighborhoods were so devastated that they became ghost towns. To house the newly homeless, the National Guard assembled tent cities in parks. School gyms were transformed into emergency shelters.

Aftershocks continued to rattle the region, setting off car alarms and jangling nerves. The night after the quake, more than 20,000 jittery southern Californians preferred to remain outdoors rather than risk sleeping inside.

National Guard troops erected a tent city in Canoga Park to house quake refugees.

They camped out in cars, on lawns, and in parks. Drawn together by the disaster, people shared food, water, and other scarce resources, even with strangers.

To prevent would-be looters from taking advantage of nightfall, the mayor declared a dusk-to-dawn curfew. Police, backed by National Guard troops in full riot gear, patrolled stricken areas.

Juan Carlo Barrio Hidalgo camped with his family on a curb near their apartment.

Food containers tumbled off the shelves of this grocery store in Moorpark, a scene repeated in markets all over the stricken area.

In Sherman Oaks, Bobby Downes tackles the ultimate kitchen chore — picking up after a quake.

As the days passed, people in the devastated neighborhoods reinvented their sense of normal. Those with homes to return to began the massive cleanup and salvaged what they could. Power companies restored service to most customers within 24 hours of the quake and to the remainder within four days. Everybody had running water by the next week. Owners of battered buildings wrestled with the decision of whether to rebuild. Most lacked earthquake insurance and would have to bear the cost of reconstruction themselves.

Many employees lost their jobs when quake-damaged businesses shut down. Workers who still had jobs faced horrendous commutes. With portions of three major freeways closed, motorists clogged the side streets. Traffic backups increased driving times by as much as two to three hours. Some commuters switched to public transportation or joined car pools.

Many children, and adults, too, were afraid to sleep. Nightmares were common. Younger children worried that the quake was punishment for something they had done wrong. Older ones fretted about the chance of worse quakes in the future. Returning to a regular routine helped most get back to normal. Although 55 percent of the 680 public school buildings in Los Angeles had been damaged, all but ten or so reopened within three weeks.

Eleven-year-old Monique Mares comforts her seven-year-old cousin, Mariella Bravo, in a tent at a North Hollywood park (*right*).

An employee of Erica's Baby Buggy store in Simi Valley clears away undamaged items (*below*).

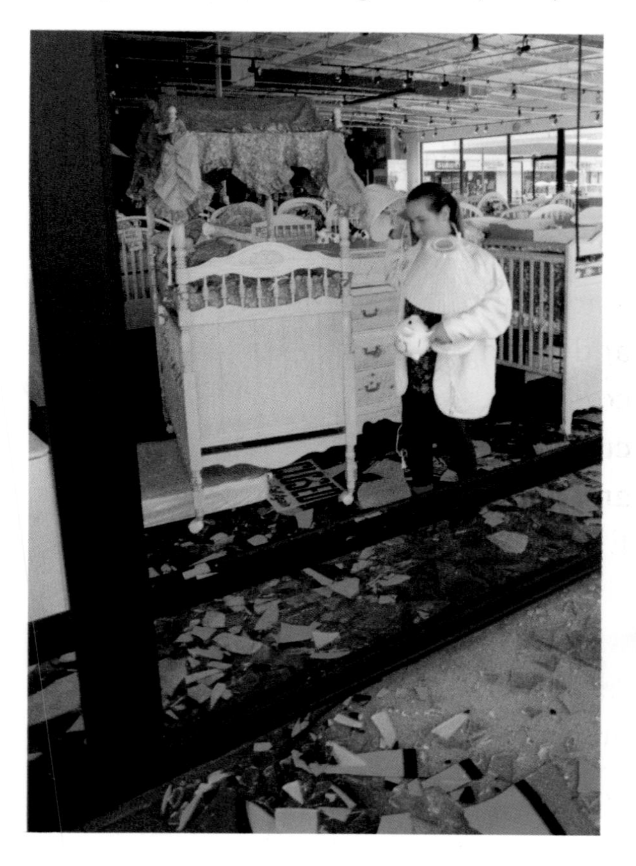

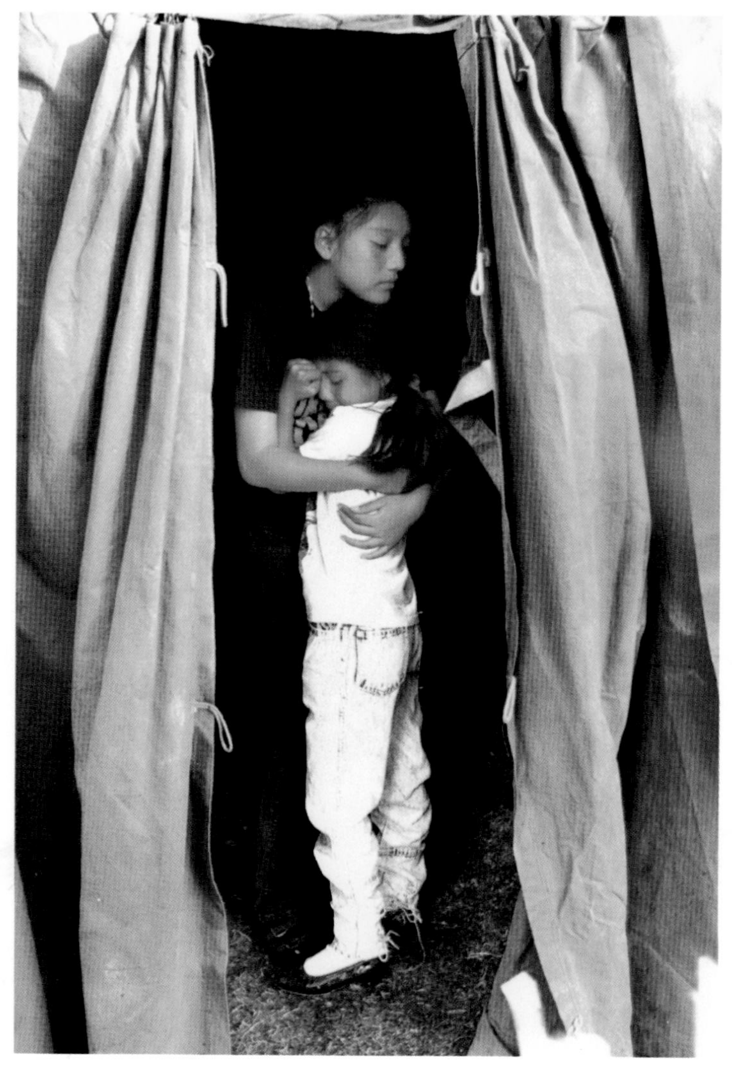

After the temblor derailed a 64-car freight train carrying sulfuric acid, emergency workers scrambled to contain the toxic chemical.

Dennis Parker stands next to an oil skimmer used to clean an oil spill in the Santa Clara River.

Repairing all the physical damage inflicted by the Northridge earthquake will take years. However, some damage required immediate attention. The cleanup of a 173,000-gallon oil spill from a burst pipeline and the safe removal of toxic chemicals at the site of a train derailment took top priority.

Road repairs began almost immediately. With nearly three million vehicles traversing Los Angeles's freeway system during rush hour, fixing the crippled freeways was a necessity. Cash bonuses were offered to contractors to finish repairs quickly. As a result, construction crews worked in shifts around the clock to beat their deadlines. Within months most repairs on major roads were completed. All were done by year's end.

Attention gradually shifted to reducing potential damage from future earthquakes. Most of the overpass collapses during the Northridge quake were caused by the shattering of shorter columns supporting the roadbed. The taller columns flexed during the quake, transferring the force of the temblor to the shorter, stiffer ones. Unable to bend as much, the shorter columns fractured. To prevent this problem from occurring again, bridges and overpasses at risk were modified, or *retrofitted.* Their short columns were "jacketed" with steel and concrete to provide extra strength and *ductility.* Ductility is the ability to stretch or bend like plastic without shattering and collapsing.

Heavy equipment claws away at the remains of a broken Simi Valley Freeway overpass (*above*).

Damage to short support columns elevating portions of freeways caused most of the overpass collapses (*left*).

When this parking garage at Cal State Northridge collapsed, its exterior reinforced concrete columns bent like plastic. However, the interior concrete columns lacked ductility, so they shattered.

A close inspection of high-rise buildings after the quake led to an unpleasant surprise: A large percentage of these supposedly quake-resistant structures had suffered cracks in the welds holding together support columns and beams. Alarmed by these findings, structural engineers immediately began to tackle the problem of how to strengthen steel-frame buildings.

The quake's devastating impact on many smaller buildings constructed on soft soil was predicted. As expected, the soft soil vibrated more violently than firmer ground. The violent shaking compacted the soil and even turned it to quicksand in the few places where sufficient groundwater was present. As the soft ground gave way, the structures on it sank and pulled apart.

To reduce destruction in future quakes, residential buildings should be strengthened by attaching plywood sheets to the exterior frames. New homes constructed on soft ground should be anchored to solid rock below the soil or designed with foundations that can float if the ground turns to quicksand. Automatic shutoff valves should be installed on natural gas lines leading to buildings.

Stress from the earthquake caused this building (*above*) in Santa Monica to pull apart along the X-shaped cracks. A new earthquake protection technique spared the headquarters of the Los Angeles County Fire Department (*below*) by reducing the amount of energy transmitted to the building during the temblor.

Preparations can be made to minimize the devastation of future quakes, but nothing can be done to prevent the temblors themselves. The reason lies in the makeup of the earth. The planet is like a roughed-up chocolate-covered cherry. The crust (the chocolate coating) is broken into about a dozen large plates and several smaller ones. The plates float on the mantle, a layer of hotter, softer rock (the gooey syrup). Beneath the mantle lies the earth's hottest layer, its dense core (the cherry).

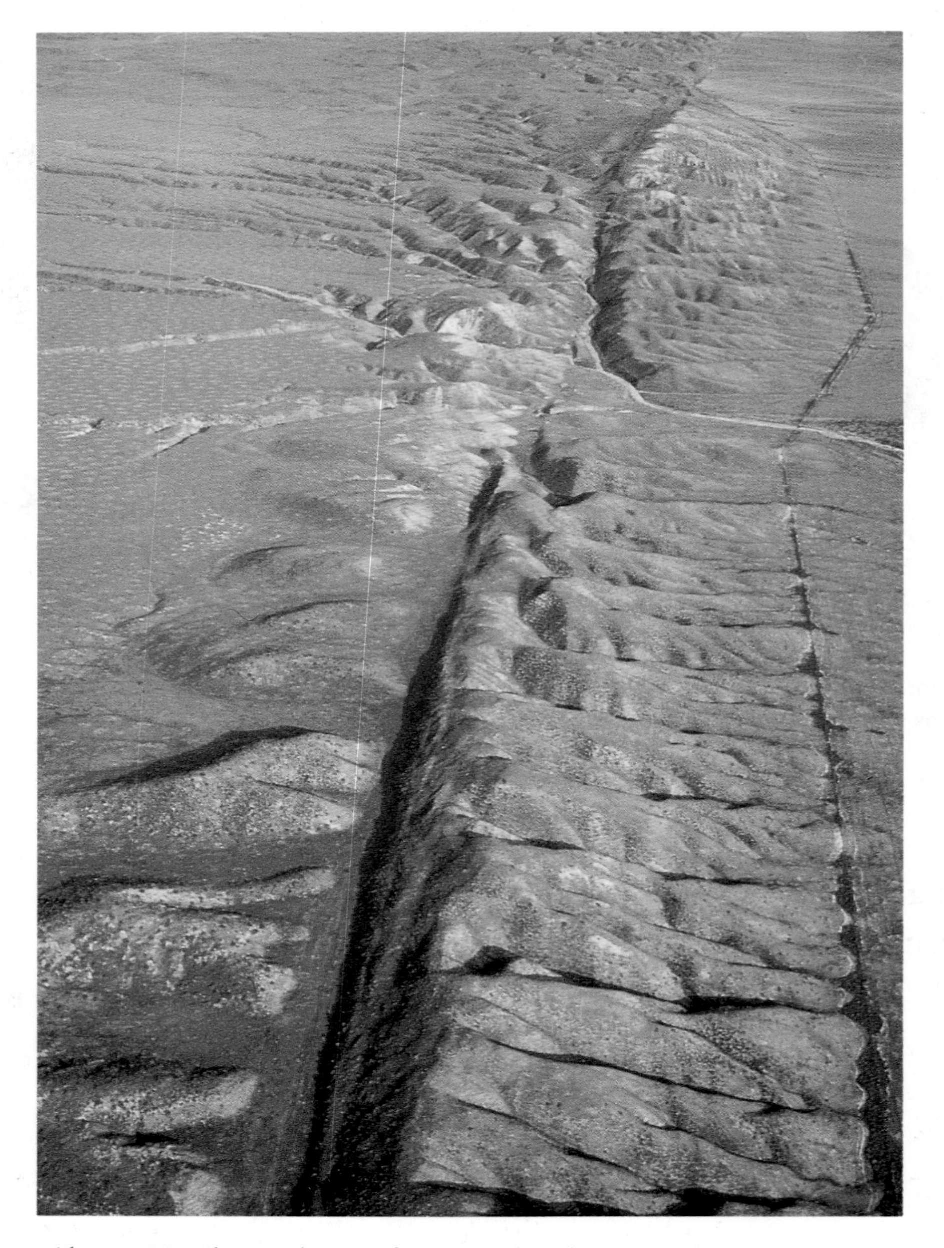

About 100 miles northwest of Los Angeles, the San Andreas Fault slices through the Carrizo Plain.

Geologist Tom Powers studies a tall scarp formed during the 1992 Landers earthquake in the Mojave Desert. Scarps are ridges that originate along a fault.

Moving relative to each other, the plates carry the continents and seafloor, which form their outer surfaces. When plates lumber past each other, their rough edges jam together. Such jamming has created the San Andreas, a *strike-slip fault,* in California. Here the North American Plate, lugging most of North America, and the Pacific Plate, hauling the Pacific Ocean and a sliver of California, grind by each other at a rate of about an inch and a half a year. In most places the two plates snag each other, preventing movement. Tremendous pressure builds up until one of the snags "snaps." Then the plates lurch past each other, the ground shifts, and the earth trembles.

The San Andreas Fault runs northwesterly for most of its length. However, north of Los Angeles, the fault twists more to the west. There the Pacific Plate no longer scrapes past the North American Plate but slams into it with titanic force. During the past few million years, this battering has splintered the crust into a vast network of *thrust faults* — ramp-shaped breaks in the earth. The battering continues to squeeze the land like an accordion. When the strain becomes too much, one of the thrust faults ruptures. As one side of the fault surges up and the other plunges down, an earthquake results. The hills and mountains of southern California were formed by such thrust faults, but until recently the faults had been relatively inactive for much of the past 200 years.

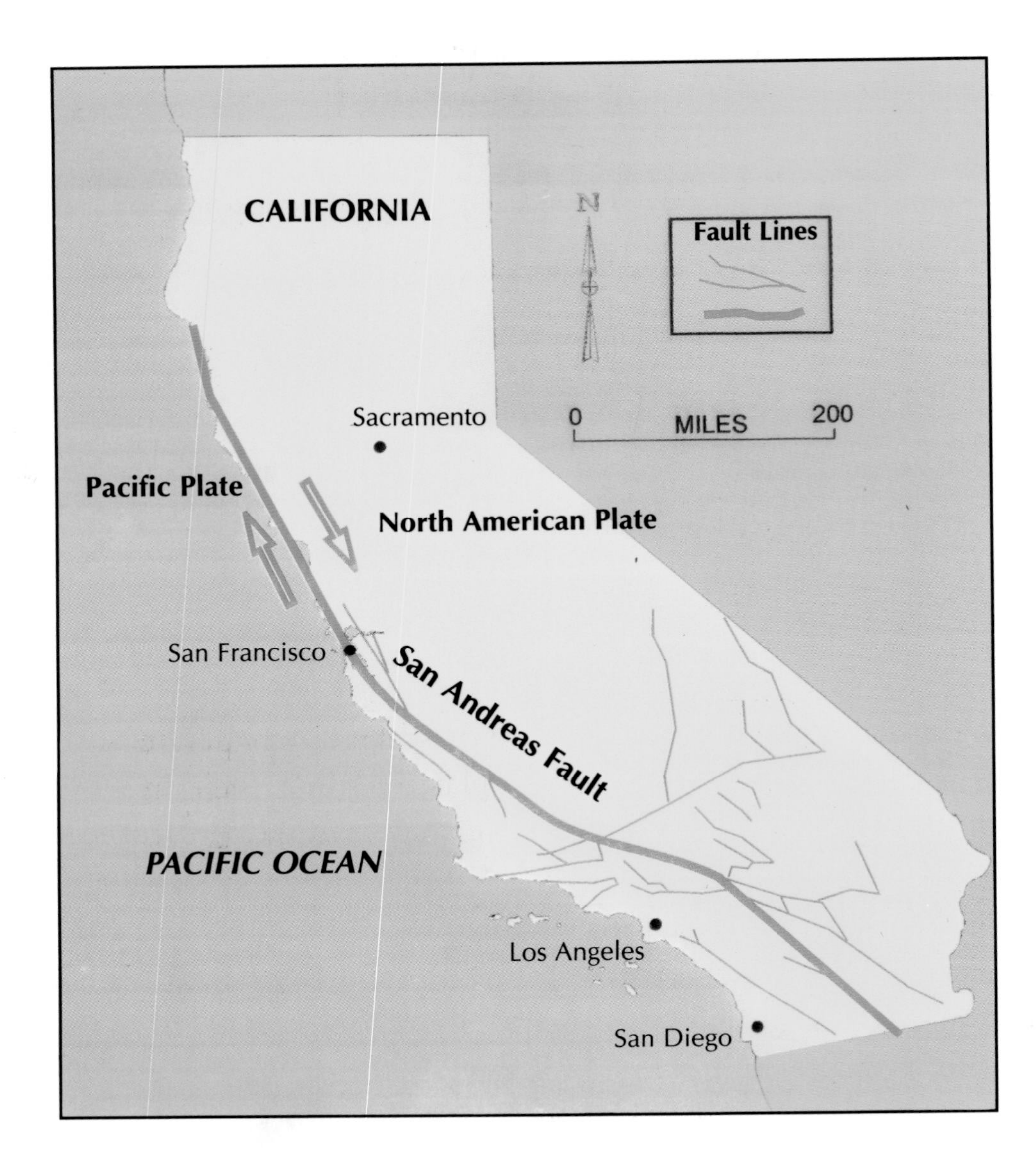

Major fault types most commonly found in the Los Angeles area

The fault responsible for the Northridge quake is a *blind thrust fault* — a fault hidden deep underground. It lies 12 miles below the surface at its southern end and rises like a ramp to a depth of three miles at the other. The 1994 rupture started at the deepest part of the fault and spread upward. In some places the underground rock slipped as much as 13 feet. The movement displaced an area 10 by 12 miles below ground but did not break the surface. However, these subsurface changes forced the overlying mountains to rise as much as 16 to 20 inches.

The Northridge earthquake was not the long-anticipated "Big One," the catastrophic quake that scientists believe will strike southern California along the San Andreas. Moderate in size, the Northridge earthquake inflicted heavy damage only because of its location beneath a densely populated region.

Recent studies suggest that the faults beneath Los Angeles may generate a big earthquake of at least magnitude seven every century or so. Such a quake could rupture the entire length of one of the six major fault zones and possibly provoke a temblor on one of the other zones. Yet no massive thrust-fault quakes have rattled Los Angeles in its 210-year history. The city may be long overdue for such a disaster.

Thrust faults formed the hills and valleys of southern California.

No one knows when or where in southern California the next monster quake will strike. But geologists are certain more temblors will come. Scientists and engineers are using data collected from the Northridge earthquake to expand their understanding of faults in the region and to improve earthquake resistance of new buildings. Federal, state, and local agencies, as well as private organizations such as hospitals and the Red Cross, are preparing their emergency response to the next big hit.

California is not the only place at risk for a major quake. The threat of earthquake looms over every state in the United States, except for large parts of Florida and Texas. At especially high risk are the western states, the Northeast, South Carolina, and the midsection of the Mississippi Valley. The earthquakes cannot be prevented. But the death toll can be minimized by revising building codes, reinforcing bridges and buildings, eliminating hazards, and training public safety organizations to respond quickly and effectively. The entire nation can learn from the devastating experience of the 1994 Northridge earthquake.